YOUR KNOWLEDGE HAS VALUE

- We will publish your bachelor's and
  master's thesis, essays and papers

- Your own eBook and book -
  sold worldwide in all relevant shops

- Earn money with each sale

Upload your text at www.GRIN.com
and publish for free

Roberto Cui

Nuclear power and energy future. A comparative analysis of Italy and Germany

GRIN Verlag

Bibliografische Information der Deutschen Nationalbibliothek:

Die Deutsche Bibliothek verzeichnet diese Publikation in der Deutschen National-
bibliografie; detaillierte bibliografische Daten sind im Internet über http://dnb.d-
nb.de/ abrufbar.

Imprint:

Copyright © 2013 GRIN Verlag GmbH
Druck und Bindung: Books on Demand GmbH, Norderstedt Germany
ISBN: 978-3-656-82208-0

This book at GRIN:

http://www.grin.com/en/e-book/282685/nuclear-power-and-energy-future-a-com-
parative-analysis-of-italy-and-germany

LLM International Energy Law and Policy

NUCLEAR POWER AND ENERGY FUTURE

ITALY AND GERMANY: A COMPARATIVE ANALYSIS

Submitted in part fulfilment of the Module

Energy Law and Policy

Roberto Cui

November 2013

Introduction

EU context and national policies

In recent times, the European institutions have been increasingly recognizing, on the one hand, the gravity of the environmental concerns related to energy (in the steps of production, consumption, and waste management and disposal), and, on the other hand, the importance of reducing the energy imports dependence of the majority of the Member countries.

These recognition has led to remarkable legislation efforts, culminated in the enactment of the so-called Third Energy Package,[1] adopted in July 2009, which, while maintaining the classic European Union approach based on liberalization and improvement of competition, also embraced the new points of view of environment, consumer protection and security of supply.[2]

As acknowledged by the EU itself, the most ambitious points of its legislation instruments are those related to the so-called 20-20-20 objectives to meet by 2020: reducing EU greenhouse gases (GHG) emissions by 20% from 1990 levels, raising the shares of EU

---

[1] The EU Third Energy Pack is formed by three Regulations (Regulation 713/2009/EC of the European Council and of the Parliament of 13 July 2009 establishing an Agency for the Co-operation of Energy Regulators – also known as ACER Regulation, Regulation 714/2009/EC of the European Parliament and of the Council of 13 July 2009 on conditions for access to the network for cross-border exchanges in electricity and repealing Regulation EC 1228/2003, Regulation 715/2009/EC of the European Parliament and of the Council of 13 July 2009 on conditions for access to the natural gas transmission networks and repealing Regulation EC 1775/2005), and two Directives (Directive 2009/72/EC of the European Parliament and of the Council of 13 July 2009 concerning common rules for the internal market in electricity and repealing Directive 2003/54/EC, and Directive 2009/73/EC of the European Parliament and of the Council of 13 July 2009 concerning common rules for the internal market in natural gas and repealing Directive 2003/55/EC).
See < http://eur-lex.europa.eu/LexUriServ/LexUriServ.do?uri=OJ:L:2009:211:0001:0014:EN:PDF>, < http://eur-lex.europa.eu/LexUriServ/LexUriServ.do?uri=OJ:L:2009:211:0015:0035:EN:PDF>, <http://eur-lex.europa.eu/LexUriServ/LexUriServ.do?uri=OJ:L:2009:211:0036:0054:en:PDF>, <http://eur-lex.europa.eu/LexUriServ/LexUriServ.do?uri=OJ:L:2009:211:0055:0093:EN:PDF>, <http://eur-lex.europa.eu/LexUriServ/LexUriServ.do?uri=OJ:L:2009:211:0094:0136:en:PDF> accessed on 4 November 2013.
[2] Angus Johnston & Guy Block, *EU Energy Law*, Oxford University Press, Oxford 2012, p. 25.

energy consumption produced from renewable sources to 20%, improving the EU's energy efficiency by 20%.[3]

The above-listed points are means to achieving a low-carbon energy structured economy for the Internal European Market. From this starting point, through a brief compared analysis between the Italian and German energy factual and legal frameworks, it is possible to outline the main critical points regarding not only the European energy policy goals, but, more broadly, the ties between energy, economics, society, and the environment.

The focus, here, is on the most recent changes affecting the Italian and German nuclear power sectors. For such purpose, after an overview of the countries' energy backgrounds (indispensable to understand the nuclear developments in the light also of the other energy resources), the likely future scenarios and the possible alternatives to meet the EU goals will be discussed, on the basis of a variety of sources.

Starting with the comparison between the Italian and German national energy balances helps to analyze the main similarities and differences between these two important European countries and to go through the current situation and seek to look to the likely future scenarios.

The Italian and German energy consumptions are articulated as the following graphs show:

---

[3] < http://ec.europa.eu/clima/policies/package/> Accessed on 1 November 2013. As it is possible to read further in the page, "the 20-20-20 targets represent an integrated approach to climate and energy policy that aims to combat climate change, increase the EU's energy efficiency and strengthen its competitiveness. [...] tackling climate and energy challenge contributes to the creation of jobs, the generation of green growth and a strengthening of Europe's competitiveness."

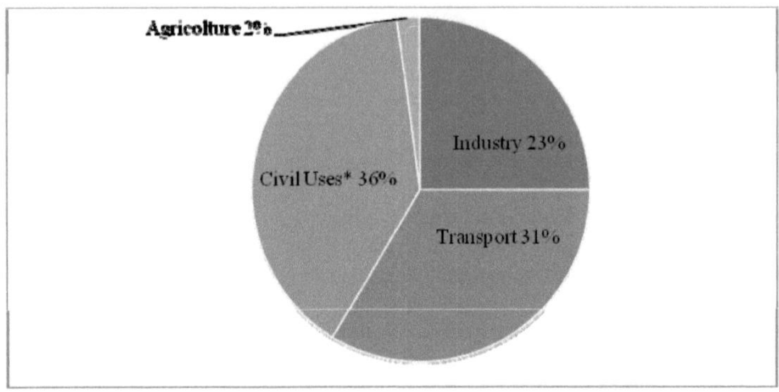

Figure 1. Italy: Final Energy Consumption by Sector (2010).[4]

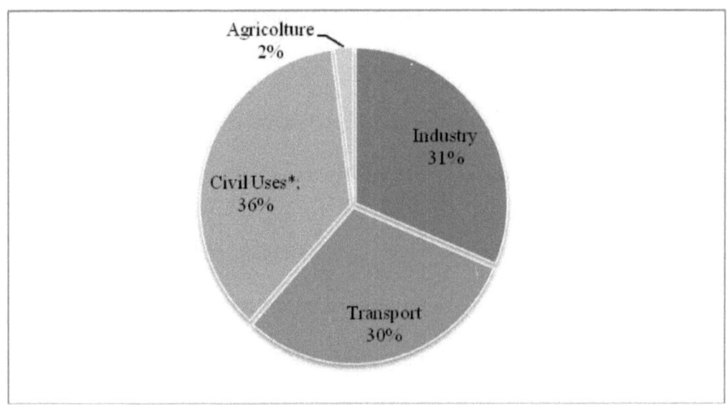

Figure 2. Germany: Final Energy Consumption by Sector (2011).[5]

---

[4] Source: adapted from <http://www.eniscuola.net/en/energy/specials/the-energy-scenario-in-italy/> (accessed on 2 November 2013 – based on the National Energy Balance published by the Ministry of Economic Development).

[5] Source: Adapted from Barbara Schlomann, Wolfgang Eichhammer, *Energy Efficiency Policies and Measures in Germany - Monitoring of EU and national energy efficiency targets*, Fraunhofer Institute for Systems and Innovation Research ISI (Germany), Karlsruhe, November 2012, as available at the following URL: <http://www.isi.fraunhofer.de/isi-media/docs/x/de/publikationen/National-Report_Germany_November-2012.pdf> Accessed on 2 November 2013.

* Civil Uses comprise: Household, Tertiary, Others.

One of the main features of the Italian energy mix is the fact that, as acknowledged by the main national energy company, ENI (*Ente Nazionale Idrocarburi* – National Body for the Hydrocarbons), "the energy policies implemented in Italy have privileged natural gas as the primary source of energy for the civil sector and for the generation of electricity."[6] (Figure 3 clearly explains the heavy Italian dependence from natural gas for its energy needs) .

While oil is used basically for the transport sector, natural gas serves as the main source for the production of electricity for household and industry uses.

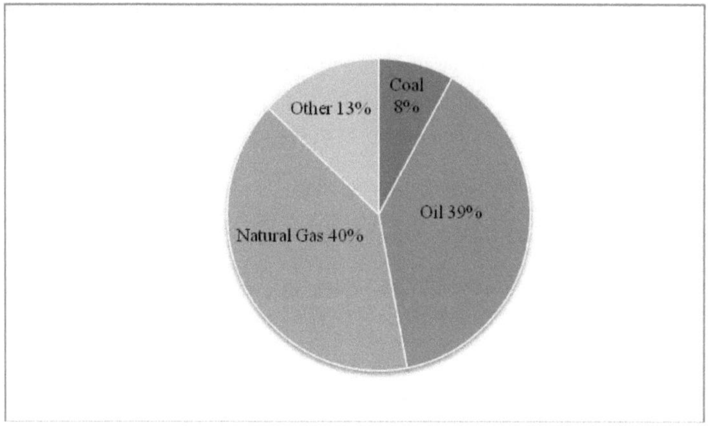

Figure 3. Italy: Primary Energy Consumption by fuel (2010 estimates).[7]

Germany relies on gas for its needs for 23% of total, as Figure 4 shows.

According to the U.S. Central Intelligence Agency (CIA), Italy is the fifth country in the world in terms of gas imports quantities, after Germany: the estimates for 2011 amounted to 70.37 and 87.57 billion cubic metres (bcm) respectively.[8]

---

[6] Benedetta Palazzo, <http://www.eniscuola.net/en/energy/specials/the-energy-scenario-in-italy/>.
[7] EnergyDelta Institute. <http://www.energydelta.org/mainmenu/energy-knowledge/country-gas-profiles/country-profile-italy#t42874> (revisited; accessed on 2 November 2013.
As one can see, Italy's electricity production from nuclear power is 0%. Italy imports all the electricity produced from nuclear power (10% of the total consumed electricity). See Nuclear Energy Agency at the following URL: <http://www.world-nuclear.org/info/Country-Profiles/Countries-G-N/Italy/> accessed on 7 November 2013.

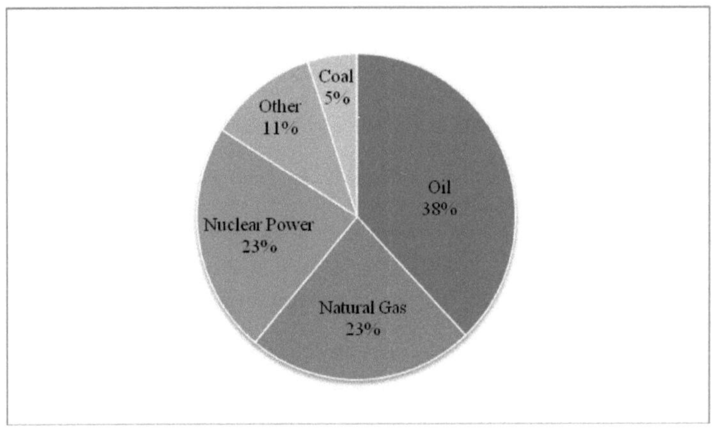

Figure 4. Germany: Electricity production by type of fuel.[9]

Italy imports most of the natural gas from few countries: Algeria (36.65%), Russia (19.65%), Lybia (12.59%), Netherlands (4.20%), Norway (3.80%).[10] Germany imported 39% of its gas from Russia in 2010, 35% from Norway, and 22% from the Netherlands.[11] This overwhelming dependence on extra EU gas represents a major concern both for the EU and for the other economic partners, such as the U.S. Indeed, as a recent interesting report prepared by the Federation of American Scientists for the U.S. Congressional Research Service states:

---

[8] See < https://www.cia.gov/library/publications/the-world-factbook/geos/it.html> and <https://www.cia.gov/library/publications/the-world-factbook/geos/gm.html> both accessed on 2 November 2013.

[9] Adapted from *Germany, Country Analysis Note*, Energy Information Administration. <http://www.eia.gov/countries/country-data.cfm?fips=GM&trk=m> Accessed on 2 November 2013.

[10] See <http://www.energydelta.org/mainmenu/energy-knowledge/country-gas-profiles/country-profile-italy#t42874> (revisited; accessed on 2 November 2013). Moreover, as further information in the same page show, the majority of these imports are made by pipelines (88%), while the Liquefied Natural Gas (LNG) imports amount to the 12%.
As regards oil, Italy is the seventh country in the world in terms of imported volumes, with 1.591 million bbl/day in the 2010 estimates (https://www.cia.gov/library/publications/the-world-factbook/geos/it.html).

[11] International Energy Agency, *Oil & Gas Security, Emergency Response of IEA Countries*, OECD – IEA, Paris 2012, as available at the following URL:
< http://www.iea.org/publications/freepublications/publication/GermanyOSS.pdf> Accessed on 2 November 2013.

5

The 28 member-state EU [including Croatia] has been a growing natural gas consumer and importer for decades. As Europe's natural gas production has declined in recent years, its dependence on imported natural gas has increased. This has left it more dependent as a whole on its primary supplier, Russia, which has shown some inclination to use its resources for political ends. Natural gas, unlike oil, which is a global commodity, is a regional commodity with regional buyers and sellers exerting more influence.[12]

The document goes on observing that "both Germany and Italy, the largest importers of Russian natural gas, have negotiated long-term deals with Russia to lock in future natural gas supplies."[13]

This few sentences clarify the American concerns about the European security of energy supply, which has been set as a major EU policy goal.[14] Furthermore, the persisting superabundant use of fossil fuels to meet the domestic energy needs is running against the achieving of the 20-20-20 European targets. In 2010, the Italian GHG records were above the EU average, while Germany was slightly below; both countries are below the European average as regards the use of renewable sources,[15] although the national estimates forecast a complete fulfilment of the European requests.[16]

---

[12] Paul Belkin, Jim Nichol, Steven Woehrel, coordinated by Michael Ratner, *Europe's Energy Security: Options and Challenges to Natural Gas Supply Diversification*, Congressional Research Service, Washington, 20 August 2013, p. 1, as available at the following URL: < http://www.fas.org/sgp/crs/row/R42405.pdf > Accessed on 1 November 2013.

[13] Ibid., p. 7.

[14] One of the most interesting point is decreasing consumption without dropping the standards of life. See, for example, European Commission, *Sustainable, secure and affordable energy for Europeans*, Publications of the European Union, Luxembourg 2012, as available at the following URL: <http://europa.eu/pol/ener/flipbook/en/files/energy.pdf> Accessed on 2 November 2013.

[15] See <http://ec.europa.eu/europe2020/europe-2020-in-your-country/italia/progress-towards-2020-targets/index_en.htm> accessed on 2 November 2013.

[16] See, for example, Strategia Energetica Nazionale – Per un'Energia più Competitiva e Sostenibile (*Italian National Energy Strategy – For a more competitive and Sustainable Energy*), published by the Ministry of the Economic Development: "In terms of energy mix, it is likely to be a renewable sources 19-20% share on the total final consumption [by 2020]." P. 5. Available at the following URL:

1. Greenhouse Gases emissions

There is no doubt that using nuclear power to produce heating and electricity is one of the possible ways to reduce greenhouse gases emissions produced by the fossil fuels. According to the U.S. Energy Information Administration, producing 1 Btu of energy by coal generates an average of 215,5 pounds of CO_2 (approximately 96 kilograms); 1 Btu of energy obtained from gasoline produces 157.2 pounds of CO_2 (about 71 kg); propane generates 139 pounds (62,5 kg) of CO_2 per Btu; natural gas emissions of CO_2 are lower but still relevant, with 117 pounds (53 kg) per Btu.[17] As regards natural gas, which, as stated above, is the most common energy resource for the production of electricity and heating in Italy and Germany (here besides nuclear power), it also produces nitrogen oxides and methane when the gas is not completely burned. Methane can also be emitted as a result of leaks and losses during transportation.[18] However, in spite of the fact that, as is has been argued, "nuclear power once promised to be the safe, clean, cheap and abundant energy source for the future,"[19] it is not free from criticisms as regards its environmental impacts.

The recent developments in Italy show the need for a resurgence of nuclear energy in the country to fight against the economic crisis (a new law on urgent provisions for economic development etc. was passed in 2008).[20] The Law No. 99 of 23 July 2009 is the first law providing for the construction of new nuclear power plants in Italy since decades, as well as

---

<http://www.sviluppoeconomico.gov.it/images/stories/normativa/20130314_Strategia_Energetica_Nazionale.pdf> accessed on 29 October 2013.

[17] See < http://www.eia.gov/tools/faqs/faq.cfm?id=73&t=11> Accessed on 2 November 2013.

[18] U.S. Environmental Protection Agency. See: <http://www.epa.gov/cleanenergy/energy-and-you/affect/natural-gas.html>. Accessed on 3 November 2013. Furthermore, "the process of extraction, treatment, and transport of the natural gas to the power plant generates additional emissions."

[19] Joseph P. Tomain, Richard Cudahy, *Energy Law in a Nutshell*, West, United States 2012, p. 430.

[20] Law No. 133 of 6 August 2008. See OECD, *Nuclear Legislation in OECD Countries, Regulatory and Institutional Framework for Nuclear Activities - Italy*, 2010, p. 5. Available at the following URL: < http://www.oecd-nea.org/law/legislation/italy.pdf> Accessed on 2 November 2013.

the first nuclear legislation instrument since the 1987 moratorium on Italy's nuclear programme.[21]

Decision-makers have to deal with the problems arising from the environmental and public health concerns in setting the future energy scenario, inasmuch as nuclear power does produce waste: notwithstanding the fact that "nuclear power plants do not emit carbon dioxide, sulphur dioxide, or nitrogen oxides as part of the power generation process,"[22] the production of electricity through nuclear power is capable to produce large amounts of water pollution substances, such as heavy metals and salts. As acknowledged by the U.S. Environmental Protection Agency, "these water pollutants, as well as the higher temperature of the water discharged from the power plant, can negatively affect water quality and aquatic life. Nuclear power plants sometimes discharge small amounts of tritium and other radioactive elements as allowed by their individual wastewater permits."[23]

The major problem related to nuclear power is the radioactivity of the wastes. As clearly explained by Tomain and Cudahy:

> The nuclear fuel cycle, from mining through reprocessing, produces four major types of waste: high level, low level, mill tailings, and gaseous effluents. [...] High level wastes [...] derive from spent nuclear fuel rods generated in the processing of

---

[21] Apart from this Law, the regulation of the Italian nuclear energy sector is based basically on the Framework Act on the Peaceful Uses of Nuclear Energy (No. 1860 of 31 December 1962), which introduces a general regime based on a series of procedural requirements such as notifications and licences, the Legislative Decree No.230 of 17 March 1995 related to the safety of nuclear installations and the protection of workers and the general public against the hazards of ionising radiation arising from the peaceful uses of nuclear energy (providing, *inter alia*, for the implementation of existing Euratom Directives on radiation protection), the Legislative Decree No.241 of 26 May 2000 which amends and completes the previous Decree, taking into account the provisions of Council Directive 96/29Euratom of 13 May 1996, the Legislative Decree No. 187 of 26 May 2000, which implements Council Directive 97/43/Euratom of 30 June 1997. (Ibid.)
According to the Nuclear Energy Agency, "the government intended to have 25% of electricity supplied by nuclear power by 2030, but this prospect was rejected at a referendum in June 2011." <http://www.world-nuclear.org/info/Country-Profiles/Countries-G-N/Italy/> accessed on 7 November 2013.
[22] U.S. Environmental Protection Agency. See: <http://www.epa.gov/cleanenergy/energy-and-you/affect/nuclear.html> accessed on 4 November 2013. Further in the page, it reads as follows: "fossil fuel emissions are associated with the uranium mining and uranium enrichment process as well as the transport of the uranium fuel to and from the nuclear plant"
[23] Ibid.

the fuel and the fabrication of plutonium. These wastes are highly toxic and remain

so for hundreds of thousand of years.[24]

Following a totally opposite path if compared to Italy, Germany has recently resumed

its nuclear power phasing-out strategy (initially proposed in the late '90s),[25] also in the light

of the Fukushima disaster in March 2011. The fundamental aim of this policy is to stop using

nuclear power for electricity-generating purposes.[26]

2. Radiation Hazards

In 2011, a popular referendum held in the Italian Region of Sardinia showed a

widespread rejection to the proposed construction of new nuclear power plants in the island.[27]

The Fukushima disaster was the main reason of the refuse, but the German example of the

Asse salt mine, in the Lower Saxony Region, can serve as an additional useful instrument to

understand the fears of a local community when a decision to build new nuclear plants is

made. Nuclear plants, as stated above, produce radioactive wastes, which have to be

temporarily stored either near the nuclear facilities or in geologically safe locations: the Asse

---

[24] Joseph P. Tomain, Richard Cudahy, Cited, p. 452.

[25] The legal instrument to implement the phasing-out decision was the Act on the Structured Phase-Out of Nuclear Power for the Commercial Production of Electricity (22 April 2002). See OECD, *Nuclear Legislation in OECD Countries, Regulatory and Institutional Framework for Nuclear Activities - Germany*, 2011, p. 5. Available at the following URL: <http://www.oecd-nea.org/law/legislation/germany.pdf>. Accessed on 3 November 2013.
Before this Act, the German nuclear energy sector was regulated by the following legislation: Articles 74, No. 11a, and 87c of the Basic Law (Federal Constitution) as amended in 1959 and in 2006, Act on the Peaceful Use of Atomic Energy and Protection against its Hazards (Atomic Energy Act) of 23 December 1959. The first phasing-out policy found its legal basis in the Act on the Structured Phase-Out of Nuclear Power for the Commercial Production of Electricity (22 April 2002). (Ibid.)

[26] The cost of replacing nuclear power with renewables is estimated by the German government to amount to some € 1,000 billion (http://world-nuclear.org/info/Country-Profiles/Countries-G-N/Germany/#.UnZQI3DlNNg, accessed on 29 October 2013). This amount does not take into account the losses in terms of high-level technical expertise, developed throughout decades and hardly replaceable.
Besides the phasing-out policy, Germany is carrying forward what is known as "*Energiewende* (Energy Transition)" which consists of highly-subsidised R&D in renewable energy sources and some extra fossil fuels capacity (see World Nuclear Association, *Nuclear Power in Germany*, October 2013: <http://www.world-nuclear.org/info/Country-Profiles/Countries-G-N/Germany/#.UnZ7YXA9Iu1> accessed on 30 October 2013).

[27] See, for example, Euronews: *Sardinia says no to nuclear power*, 18 May 2011. Available at the following URL: <http://www.euronews.com/2011/05/18/sardinia-says-no-to-nuclear-power/> Accessed on 2 November 2013.

salt mine was chosen in 1967 as the perfect place to store the German nuclear industry wastes. Yet, approximately ten years later, it was found that the Southern part of the mine – which was the one chosen for the storage – is likely to be involved in infiltration from the aquifers. Water infiltrations in the mine represent a serious danger inasmuch as there is a high likelihood of corrosion of the nuclear wastes drums.[28]

As asserted by the World Health Organization (WHO), "nuclear power stations, waste treatment plants and storage areas all present problems, and it is clear that their seriousness depends to a large extent on the potential risks of the facilities concerned [...] and their planned operating period, particularly for long-term waste storage."[29]

Thus, although the external costs of producing electricity from nuclear power are estimated to be lower than the methods based on other fuels, such as coal, oil or gas, to the aim of a careful evaluation of the possible benefits and risks involved in the decision to build new nuclear plants, it has at least to be note that:

> The results of assessments like these are interesting, but are fraught with methodological problems, for example concerning how to calculate the cost of specific types of damage. Simple economic assessment, based on insurance replacement costs for example, may not provide a realistic measure of, or proxies for, the human value of amenity loss or health damage, much less the ecological alue of any disruption. Even more contentious is the value put on human life [...].[30]

---

[28] See:< http://www.fraktion.gruene-niedersachsen.de/themen/themenspecial-endlager/artikel/radioactive-waste-in-the-asse-salt-mine-in-germany.html> and
<http://www.asse.bund.de/EN/2_WhatIs/RadioactiveWaste/_node.html> both accessed on 5 November 2013.
[29] World Health Organization – Europe, *Nuclear Power and Health, The Implications for health of nuclear power production*, WHO Regional Publications, European Series No. 51, p. 29. Further in the same publication, it reads as follows: "Unlike the release of effluents into surface water or the atmosphere which, except for accidents, forms part of normal procedures within authorized limits, release into the soil is illegal in most countries." (Ibid.)
[30] Godfrey Boyle, Bob Everett, Janet Ramage, *Energy Systems and Sustainability, Power for a Sustainable Future*, Oxford University Press, United Kingdom 2003, p. 564.

Nonetheless:

> [...] even though the absolute magnitudes of the external costs of energy sources
> have proved to be highly-sensitive to changing assumptions and research results,
> the overall conclusions on the *relative* ranking of energy sources in terms of their
> external costs have not altered.[31]

3. Beyond nuclear: similarities in opposite energy policies

Italy and Germany have almost the same amount of recoverable uranium resources: 4,800 tonnes each as regards 130 U.S. Dollars (USD)/ kg uranium, 6,100 and 7,000 tonnes respectively as regards 260 USD/kg uranium.[32] As paradoxically recognized by the Italian Ministry of Economic Development itself, "[The use of nuclear power] is expected to increase only in the non-OECD countries, in particular in China, India, Korea and Russia, whereas meaningful developments are unlikely to be registered in the Western countries (especially in Europe), because of a high costs/risks economic profile, and because of the apprehensions caused by the current technologies, which will bring to a re-evaluation of the safety limits of the operating or under construction plants and to a renewed effort by the Western countries on research, reduction and disposal of the wastes, and international cooperation for the safe use in civil uses."[33] At the same time, Italy embraces the EU Commission Energy Roadmap 2050, which forecasts a GHG emissions reductions by 80-95% within 2050 compared to the 1990 levels. To do so, the EU stresses the importance of the energy efficiency improvement and energy production from renewable sources, without disregarding a look to the nuclear power and the development of the Carbon Capture and

---

[31] Ibid.

[32] OECD-NEA, *Uranium 2009: Resources, Production and Demand*, 2010, p. 18. Available at the following URL: <http://www.oecd-nea.org/ndd/pubs/2010/6891-uranium-2009.pdf> accessed on 3 November 2013.

[33] *Italian National Energy Strategy*, cited, p. 8.

Storage (CCS) Technology. Moreover, the provisions of the Energy Roadmap assign a fundamental role to natural gas in the transition period.[34]

In Germany, the *Energiewende* (Energy Transition) program set a framework whereby "the withdrawal from nuclear energy is necessary and is recommended to rule out future risks that arise from nuclear [...]. It is possible because there are less risky alternatives."[35] Besides the focus on renewables – exactly as the Italian provisions – the German energy transition is supposed to occur on the basis of a continuing reliance on natural gas:

> The gap in supply caused by the withdrawal from nuclear energy is mainly to be filled by the use of renewable energy sources and improving energy efficiency as well as with the aid of fossil fuels, especially gas. They provide security of a long – term available power supply. This gap can be filled without compromising the ambitious targets and within the upper limits for greenhouse gas emissions which are legally established in the EU. Natural gas has an important role to play in this regard. [It] is the fossil fuel with the lowest CO_2 figures and is definitely available in the transitional period. Germany's dependence on gas deliveries can be counteracted by the infrastructurally-sound access to various sources of supply.[36]

---

[34] Ibid., p. 13.

[35] Ethics Commission for a Safe Energy Supply, *Germany's energy transition – A collective project for the future*, Berlin 30 May 2011, p. 4. Available at the following URL:
<http://www.bundesregierung.de/Content/DE/_Anlagen/2011/05/2011-05-30-abschlussbericht-ethikkommission_en.pdf;jsessionid=5A12D13818856C69E006AFD6DC52AB90.s1t2?__blob=publicationFile&v=2> accessed on 3 November 2013.
Reading further in the report: "The withdrawal from the use of nuclear energy offers many opportunities. Science in Germany is in an excellent position and can be relied upon to provide further significant innovative and highly-efficient solutions for the energy transition."

[36] Ibid., pp. 35-36.

With particular regard to the last statement, it is interesting to note that the German situation is similar to the Italian one also in respect to the international energy infrastructure that allow the two countries to get the resources that are needed.[37]

Conclusion

This work sought to analyse critically the energy law frameworks and policies of two EU member countries with regard to their energy mixes, in the light of their international and regional obligations and strategies as well as their domestic policies and the points of view of local communities, which have the right to counter the states decisions by asserting their rights.

As highlighted above, albeit both countries are committed to achieving the objectives set by the EU, they are following two different paths as regards their nuclear policies. Still, there are evident similarities between the Italian and German policies on natural gas and renewable sources (now omitting any judge about the outcomes) both at the domestic and at the international level.

---

[37] Exactly as Germany has given its support to the construction of the Russian North Stream, Italy – with the same carelessness in regard to the European preferences for the U.S.-supported Southern Corridor – ensured its support for the construction of the Moscow-Gazprom led South Stream pipeline. As it is possible to read in the already cited FAS Report for the Congressional Research Service, entitled *Europe's Energy Security: Options and Challenges to Natural Gas Supply Diversification*: "In response to past supply cutoffs and the potential for future energy supply interruptions, European leaders [...] have sought to increase their energy security by exploring supply diversification options. One such response, though contrary to the U.S. perspective of energy security through diversification, has been the decision by some EU members to support alternative transit routes for Russian gas. This includes Germany's decision to support construction of the Nord Stream pipeline, which directly connects Russia and Germany, Russia's largest importer."
See also Stratfor, Brief: *Russian, German Leaders Attend Nord Stream Ceremony*, 9 April 2010, as regards the recent strengthening of ties between German and Russia on energy issues. Available at the following URL: <http://www.stratfor.com/analysis/brief-russian-german-leaders-attend-nord-stream-ceremony> accessed on 3 November 2013. And – as regards the energy link between Italy and Russia – Stratfor, *Russia: Italy's Eni To Help Speed South Stream Project*, 16 January 2009, available at the following URL: <http://www.stratfor.com/situation-report/russia-italys-eni-help-speed-south-stream-project> accessed on 3 November 2013.

The supporters of nuclear power claim that it is the most safe and affordable energy source to reduce the greenhouse gases emissions, consequently reducing the air pollution and improving the health conditions of the European people without affecting their standards of life. On the other hand, there is a considerable public and scientific opinion according to which the risks of the nuclear fuel cycle (i.e. radioactive wastes, accident hazards) overcome its advantages in terms of reduction of GHG emissions.[38]

Increasing R&D funding in renewables, in order to raise the reliance on these traditional energy resources, has been identified as a solution to the problem of deciding between fossil fuels or nuclear power. In spite of this, there are many obstacles to achieve the targets set by the EU or, more in general, to meet the goal of a higher share of renewables in the national energy mixes.[39] In 2009, 11.7% of energy used in the EU came from renewable sources.[40]

---

[38] According to the Eurobarometer surveys data, "[...] 49% [of Europeans] do not consider that the disposal of radioactive waste can be carried out safely and 45% disagree that nuclear materials are sufficiently protected against malevolent use." In Italy, the percentage of people who don't think that it is possible to operate a nuclear power plant in a safe manner is 31%, while in Germany this figure is 44%.
Furthermore: "The final management of radioactive waste has been the subject of debate in many countries. Results show that it is still an issue determining perceptions of nuclear risk, with half of respondents (49%) disagreeing with the statement, compared to 40% who hold a more favourable view. [...]two countries with Nuclear Power Plants in operation record the highest levels of disbelief in the safe management of radioactive waste: Germany (70%) and France (66%)." As regards Italy, the figure is 46% of people who don't believe that the disposal of radioactive waste can be done in a safe manner, against 42% who think it can be done.
Moreover, although the majority of people in the EU consider that nuclear power helps to limit climate change, the gap between positive and negative answers is not so wide (10% for the EU-27, 8% for Italy and only 3% for Germany).
See European Commission, Special Eurobarometer 324, *Europeans and Nuclear Safety*, Brussels 2010, pp. 14, 52-4, 63.
Available at the following URL:
<http://ec.europa.eu/energy/nuclear/safety/doc/2010_eurobarometer_safety.pdf> accessed on 5 November 2013.
[39] Using the IEA wording: "Taking all new developments and policies into account, the world is still failing to put the global energy system onto a more sustainable path. Global energy demand grows by more than one-third over the period to 2035 [...]. Energy demand barely rises in OECD countries, although there is a pronounced shift away from oil, coal (and, in some countries, nuclear) towards natural gas and renewables." (International Energy Agency, *World Energy Outlook 2012*, OECD-IEA, Paris 2012, p.1. Available at the following URL: < http://www.iea.org/publications/freepublications/publication/English.pdf> accessed on 5 November 2013.)
[40] <http://www.eea.europa.eu/highlights/massive-renewable-energy-growth-this/key-facts/percentage-of-renewable-energy> Accessed on 5 November 2013.

Notwithstanding these obstacles, progresses have been made in the field of renewables, which can be a gleam of hope for the future of energy developments in Europe: by way of example, an international investment worth €300 million has led to the construction of a biorefinery in the Italian Region of Piemonte, the first able to produce biofuels from non-alimentary biomasses.[41]

---

[41] See: < http://www.reuters.com/article/2012/07/10/idUS122417+10-Jul-2012+HUG20120710> and John Acher, *Novozymes says Italy plant heralds biofuel future*, Reuters.com, 12 April 2011. Available at the following URL: <http://www.reuters.com/article/2011/04/12/us-novozymes-biofuels-plant-idUSTRE73B2KR20110412>. Both accessed on 5 November 2013.

SOURCES

<u>LEGISLATION</u>

EU

Directive 2009/72/EC of the European Parliament and of the Council of 13 July 2009 concerning common rules for the internal market in electricity and repealing Directive 2003/54/EC.

Directive 2009/73/EC of the European Parliament and of the Council of 13 July 2009 concerning common rules for the internal market in natural gas and repealing Directive 2003/55/EC).

Regulation 713/2009/EC of the European Council and of the Parliament of 13 July 2009 establishing an Agency for the Co-operation of Energy Regulators – also known as ACER Regulation.

Regulation 714/2009/EC of the European Parliament and of the Council of 13 July 2009 on conditions for access to the network for cross-border exchanges in electricity and repealing Regulation EC 1228/2003.

Regulation 715/2009/EC of the European Parliament and of the Council of 13 July 2009 on conditions for access to the natural gas transmission networks and repealing Regulation EC 1775/2005.

ITALY

Framework Act on the Peaceful Uses of Nuclear Energy (No. 1860 of 31 December 1962).

Law No. 133 of 6 August 2008. (Urgent provisions for economic development, simplification, competitiveness, stabilization of public finance and tax equalization).

Law No. 99 of 23 July 2009. (Provisions for the development and internationalization of enterprises, as well as on energy).

Legislative Decree No.230 of 17 March 1995. (Implementation of Directives 89/618/Euratom, 90/641/Euratom, 92/3/Euratom and 96/29/Euratom on ionising radiation).

Legislative Decree No. 187 of 26 May 2000. (Implementation of Directive 97/43 / Euratom on health protection of individuals against the dangers of ionizing radiation in relation to medical exposures).

Legislative Decree No.241 of 26 May 2000. (Implementation of Directive 96/29/EURATOM for the protection of the general public and workers against the dangers arising from ionizing radiation).

GERMANY

Act on the Peaceful Use of Atomic Energy and Protection against its Hazards (Atomic Energy Act) of 23 December 1959.

Act on the Structured Phase-Out of Nuclear Power for the Commercial Production of Electricity (22 April 2002).

Act on the Structured Phase-Out of Nuclear Power for the Commercial Production of Electricity (22 April 2002).

Articles 74, No. 11a, and 87c of the Basic Law (Federal Constitution) as amended in 1959 and in 2006.

<u>OFFICIAL PUBLICATIONS</u>

Congressional Research Service (Paul Belkin, Jim Nichol, Steven Woehrel, coordinated by Michael Ratner), *Europe's Energy Security: Options and Challenges to Natural Gas Supply Diversification*, Congressional Research Service, Washington, 20 August 2013.

Ethics Commission for a Safe Energy Supply, *Germany's energy transition – A collective project for the future*, Berlin 30 May 2011.

European Commission, Special Eurobarometer 324, *Europeans and Nuclear Safety*, Brussels 2010.

European Commission, *Sustainable, secure and affordable energy for Europeans*, Publications of the European Union, Luxembourg 2012.

Fraunhofer Institute for Systems and Innovation Research ISI (Germany, Barbara Schlomann, Wolfgang Eichhammer), *Energy Efficiency Policies and Measures in Germany - Monitoring of EU and national energy efficiency targets*, , Karlsruhe, November 2012.

International Energy Agency, *Oil & Gas Security, Emergency Response of IEA Countries*, OECD – IEA, Paris 2012.

International Energy Agency, *World Energy Outlook 2012*, OECD-IEA, Paris 2012.

Ministero dello Sviluppo Economico, *Strategia Energetica Nazionale - Per un'Energia più Competitiva e Sostenibile* (Italian Ministry for the Economic Development, *National Energy Strategy - For a more Competitive and Sustainable Energy*), Rome 2012.

OECD, *Nuclear Legislation in OECD Countries, Regulatory and Institutional Framework for Nuclear Activities - Germany*, 2011.

OECD, *Nuclear Legislation in OECD Countries, Regulatory and Institutional Framework for Nuclear Activities - Italy*, 2010.

OECD-NEA, *Uranium 2009: Resources, Production and Demand*, 2010.

World Health Organization – Europe, *Nuclear Power and Health, The Implications for health of nuclear power production*, WHO Regional Publications, European Series No. 51.

BOOKS

Godfrey Boyle, Bob Everett, Janet Ramage, *Energy Systems and Sustainability, Power for a Sustainable Future*, Oxford University Press, United Kingdom 2003.

Joseph P. Tomain, Richard Cudahy, *Energy Law in a Nutshell*, West, United States 2012.

Angus Johnston & Guy Block, *EU Energy Law*, Oxford University Press, Oxford 2012.

asse.bund.de

bundesregierung.de

cia.gov

ec.europa.eu

eea.europa.eu

eia.gov

energydelta.org

eniscuola.net

epa.gov

eur-lex.europa.eu

euronews.com

europa.eu

fas.org

fraktion.gruene-niedersachsen.de

iea.org

isi.fraunhofer.de

oecd-nea.org

reuters.com

stratfor.com

sviluppoeconomico.gov.it

world-nuclear.org